Saxon MATH™

Intermediate 5

Instructional Masters

Stephen Hake

SAXON™

A Harcourt Achieve Imprint

www.SaxonPublishers.com
1-800-284-7019

Instructional Masters

Saxon Math Intermediate 5 Instructional Masters contains a Parent Letter, Lesson Recording Forms, daily Power-Up activities, a Student Progress Chart, Test-Day Activities, and Lesson Activities. Brief descriptions of these components are provided below.

Parent Letter

Parental involvement greatly improves a student's chance for success. When parents understand what is being asked of their children in school, they can provide support at home. The Parent Letter explains the keys to success in the Saxon Mathematics program. An English and a Spanish version are provided. At the beginning of the year, give each student one copy to take to his or her parents. To ensure that the parents receive the letters, you might require the students to return the letters bearing their parents' signatures. Alternatively, you could mail the letters directly to the parents.

Recording Forms

There are two kinds of recording forms you might find useful. The Lesson Recording Forms are designed to help you track and analyze student performance on lessons. The Student Progress Chart is designed to help students track and analyze their daily Facts practice performance.

Answer Form A: Lesson Worksheet

This is a single-page master with check boxes for daily activities, answer blanks for Power-Up activities, and partitions for recording the solutions to Lesson Practice problems.

Answer Form B: Written Practice Solutions

This is a double-page master with a grid background and partitions for recording the solutions to 30 Written Practice problems.

Answer Form C: Written Practice Solutions

This is a double-page master with a plain, white background and partitions for recording the solutions to 30 Written Practice problems.

Form D: Student Progress Chart

This is a single-page master with boxes for recording the time and score of each Power-Up Facts practice. Give one to each student so he or she can monitor Facts practice performance throughout the year.

Power Up

Each instructional day begins with a Power-Up activity which should be administered before every lesson. The Power Up is an integral part of the daily program. Each Power Up contains a set of math Facts, eight Mental Math problems, and a Problem Solving activity.

Facts

Students are expected to master a variety of mathematical facts, skills, and vocabulary. Daily Facts practice enhances the retention of this learning. Facts practice covers content that students should be able to recall immediately. Rapid and accurate recall of basic facts, skills, and vocabulary dramatically increases students' mathematical abilities. Mastery of basic facts frees students to focus on procedures and concepts rather than computation. Employing memory to recall frequently encountered facts permits students to bring higher-level thinking skills to bear when solving problems.

Reproduce the Power Up indicated for each lesson and provide a copy to each student.* On the first presentation of a Facts practice set, you may choose to make the practice instructional by working through the problems with the students and providing the answers. However, students should work independently and rapidly during all subsequent practices of the same Facts sheet, striving to improve on previous performances.

Mental Math

The Mental Math problem set follows Facts practice. Students should perform calculations mentally without the aid of pencil and paper or the use of a calculator. Mental math ability pays lifelong benefits and improves with practice.

At the conclusion of Facts practice, begin the Mental Math problem set. Allow two or three minutes to complete the Mental Math portion of each Power Up. Then review the correct answers with the class. Instruct students to correct any mistakes.

Problem Solving

A Problem Solving activity follows Mental Math. Each problem provides students the opportunity to develop and practice problem-solving strategies using a four-step problem-solving process to communicate mathematical ideas. With daily practice, students can become powerful problem solvers.

The problem-solving exercise should be approached as a whole-class activity, providing a rich problem-solving experience for all students.

---

* As an alternative to copying a Power Up for each student every day, provide each student with a Saxon Math Intermediate 5 *Power-Up Workbook*, which contains all the Power Ups required for one student for one year.

Encourage students to propose a problem-solving strategy. Then, as a group, apply the four-step problem-solving process to the problem. Instruct students to record the problem-solving strategy and the solution to the problem on their Power-Up worksheet.

Most problem-solving exercises can be solved in a few minutes. The entire Power Up should take less than 15 minutes to complete.

Each Power Up contains a line for students to record their Facts practice times. Students should also record their time and score on a copy of the Student Progress Chart (Form D). Timing students is motivating. Striving to improve speed helps to automate skills and offers the additional benefit of an up-tempo atmosphere for the class. Time invested in Facts practice is repaid in the students' ability to work faster. Allow five minutes or less for Facts practice.

Quickly read answers at the conclusion of the Facts practice. The answers are found at the bottom of the lesson page in the *Teacher's Manual* or in the Answer Key of this book. Students should correct errors and complete the sheet as part of the day's assignment if they are unable to finish within the allotted time.

Test-Day Activity Masters

A Test-Day Activity is a brief exercise that utilizes class time that remains after students complete specific assessment tests. The *Teacher's Manual* recommends one activity for use after each even-numbered Cumulative Test. Each Test-Day Activity is supported by a worksheet for which masters are included in this book. Copy the required master and distribute the worksheet to students.

Lesson Activity Masters

Selected lessons and investigations in the student textbook present content through activities. Often these activities require the use of Lesson Activity masters. These Lesson Activity masters should be photocopied for students to use during the appropriate lesson or investigation. Although the Lesson Activity masters support specific lessons and investigations, they can also be used throughout the school year to complement instruction and review.

We suggest that the fraction manipulatives (Lesson Activities 24–29) be photocopied on different colors of paper or that students color-code their copies of the fraction manipulatives before cutting them. Some of these masters are worksheets. Answers to these worksheets can be found in the Answer Key of this book.

Dear Parent or Guardian:

This year your child will be using a **Saxon Math** textbook. **Saxon** books are unlike traditional texts in which students are taught and expected to learn an entire mathematical concept in one day. Instead of practicing 30 problems of the same kind, **Saxon Math** builds mastery through daily practice of skills and concepts taught in earlier lessons.

Saxon textbooks divide concepts into small, easily grasped pieces called increments. A new increment is presented each day, and students work only a few problems each day involving the new material. The remaining homework problems cover previously introduced concepts. Thus, every assignment (and every test) is a review of material covered up to that point, providing students with the time and opportunity to understand and remember necessary concepts.

This is where your involvement is crucial. *It is essential that your child complete each day's assignment.* You can expect your child to complete three to five lessons per week with a test after every fifth lesson. Please ask to see your child's daily assignment, and encourage him or her to complete all the work. Students who complete all the daily assignments will have much to be proud of at test time, so you will have much to praise.

Some key words in the preface of your child's textbook are printed in boldface. They remind your child of the importance of the daily exercises. Those words are worth repeating here:

> *Solving these problems day after day is the secret to success. Work every problem in each Lesson Practice, Written Practice, and investigation. Do your best work and you will experience success and true learning, which will stay with you and serve you well in the future.*

Learning mathematics is similar to learning a foreign language, a musical instrument, or an athletic skill. Long-term practice is necessary to develop and maintain mastery of the problem-solving skills needed for success in math-intensive disciplines.

Your child's experience in mathematics will be a direct result of his or her attitude and effort. A positive attitude and a commitment to long-term practice greatly improve the chances of your child progressing into higher-level mathematics courses and experiencing success in other fields that require mathematical understanding. Through our combined efforts your child can experience success at this level and learn the foundational skills needed for the next level.

Sincerely,

Estimados padres o tutores:

Este año su hijo(a) usará el libro de texto *Saxon* de matemáticas. Los libros de *Saxon* no son como los textos tradicionales, los cuales enseñan y esperan que los estudiantes aprendan todo un concepto matemático en un sólo día. En lugar de practicar 30 problemas del mismo tipo, las matemáticas de *Saxon* desarrollan destreza a través de la práctica de habilidades y conceptos enseñados en clases previas.

Los libros de texto de *Saxon* dividen los conceptos en unidades más pequeñas y más fáciles de comprender, llamadas incrementos. Un incremento nuevo es introducido cada día y diariamente los estudiantes resuelven sólo unos cuantos problemas que tratan sobre el material nuevo. El resto de los problemas de tarea cubren conceptos presentados previamente. De esta forma, cada tarea (y cada examen) es un repaso del material ya cubierto, lo cual da tiempo a los estudiantes y les ofrece la oportunidad de comprender y recordar conceptos necesarios.

Aquí es donde la participación de usted es crítica. *Es esencial que su hijo(a) termine la tarea de cada día.* Puede contar con que su hijo(a) cubrirá de 3 a 5 lecciones por semana y tomará un examen después de cada cinco lecciones. Por favor, pida a su hijo(a) que le muestre la tarea todos los días y aliéntelo(a) a que la haga en su totalidad. Los estudiantes que hayan terminado todas las tareas diarias se sentirán muy orgullosos del resultado de sus exámenes y ustedes tendrán mucha razón para felicitarlos.

Unas palabras claves del prólogo del libro de su hijo(a) están impresas en letra negrilla. Sirven para recordar a su hijo(a) la importancia de los ejercicios diarios. Vale la pena repetir esas palabras aquí:

> *Solucionar estos problemas día a día es el secreto del éxito. Resuelve cada problema en cada sección de práctica, en cada sección de problemas y en cada investigación. No te saltes problemas. A través del esfuerzo honesto obtendrás éxito y aprendizaje verdadero que mantendrás toda la vida y que te será muy útil en el futuro.*

El aprendizaje de matemáticas es similar al de un idioma extranjero, un instrumento musical, o una habilidad atlética. El llegar a ser competente requiere práctica a largo plazo para desarrollar y mantener el dominio de las destrezas necesarias para tener éxito en disciplinas que requieren muchas matemáticas.

La experiencia de su hijo(a) en matemáticas será el resultado directo de su actitud y su esfuerzo. Una actitud positiva y un compromiso a la práctica a largo plazo incrementan considerablemente la posibilidad de que su hijo(a) progrese hacia los cursos de matemáticas avanzadas y que tenga éxito en otras disciplinas que requieren compresión de matemáticas. A través de la combinación de nuestros esfuerzos, su hijo(a) puede lograr el éxito en este nivel y aprender los conocimientos fundamentales que necesitará para el siguiente nivel.

Sinceramente,

Name _____

Lesson _____

Show all necessary work.
Please be neat.

Power Up
☐ Facts
☐ Mental Math
☐ Problem Solving

Review
☐ Homework Check
☐ Error Correction

Instruction
☐ New Concept
☐ Lesson Practice
☐ Written Practice

Facts

Test:	Time:	Score:

Mental Math

a.	b.	c.	d.
e.	f.	g.	h.

Problem Solving Strategies   Check any you use.

☐ Act it out or make a model	
☐ Use logical reasoning	
☐ Draw a picture or diagram	
☐ Write a number sentence or equation	
☐ Make it simpler	
☐ Find/Extend a pattern	
☐ Make an organized list	
☐ Guess and check	
☐ Make or use a table, chart or graph	
☐ Work backwards	

Lesson Practice

a.	b.	c.
d.	e.	f.
g.	h.	i.
j.	k.	l.

Name _____

Lesson _____

Show all necessary work. Please be neat.

1.	2.	3.

4.	5.	6.

7.	8.	9.

10.	11.	12.

13.	14.	15.

 Saxon Math Intermediate 5

16.	17.	18.
19.	20.	21.
22.	23.	24.
25.	26.	27.
28.	29.	30.

Name _____

Lesson _____

Show all necessary work. Please be neat.

1.	2.	3.
4.	5.	6.
7.	8.	9.
10.	11.	12.
13.	14.	15.

Saxon Math Intermediate 5

16.	17.	18.
19.	20.	21.
22.	23.	24.
25.	26.	27.
28.	29.	30.

Name _____

Power Up Facts	# Possible	Time and Score time / # correct							
A Addition Facts	40								
B Subtraction Facts	40								
C Multiplication Facts	40								
D Division Facts	40								
E Division Facts	40								
F Multiplication Facts	40								
G Division with Remainder	20								
H Improper Fractions	20								
I Fractions to Reduce	20								
J Mixed Numbers	20								
K Percents	16								

Saxon Math Intermediate 5

Name _____ Time _____

Facts   Add.

5 +5	2 +9	4 +5	3 +7	8 +8	2 +6	6 +9	4 +8	2 +4	7 +9
3 +4	7 +8	5 +9	2 +3	4 +9	6 +6	5 +0	3 +8	10 +10	5 +6
0 +0	2 +7	9 +9	5 +7	3 +3	4 +6	2 +2	9 +1	8 +9	3 +6
4 +4	3 +9	2 +5	6 +8	7 +7	3 +5	5 +8	4 +7	2 +8	6 +7

Mental Math

a.	b.	c.	d.
e.	f.	g.	h.

Problem Solving

Understand

What information am I given?

What am I asked to find or do?

- -

Plan

How can I use the information I am given?

Which strategy should I try?

- -

Solve

Did I follow the plan?

Did I show my work?

Did I write the answer?

- -

Check

Did I use the correct information?

Did I do what was asked?

Is my answer reasonable?

Saxon Math Intermediate 5

Facts   Subtract.

9 −8	8 −5	16 − 9	11 − 9	9 − 3	12 − 4	14 − 9	6 −4	16 − 8	5 −2
14 − 7	20 −10	10 − 7	15 − 6	13 − 7	18 − 9	10 − 8	7 −3	11 − 5	9 −4
12 − 6	10 − 5	17 − 9	13 − 8	12 − 3	7 −2	14 − 8	8 −6	15 − 7	13 − 9
8 −4	12 − 5	9 −2	16 − 7	11 − 8	6 −3	10 − 6	17 − 8	10 −10	11 − 4

Mental Math

a.	b.	c.	d.
e.	f.	g.	h.

Problem Solving

(Understand)

What information am I given?

What am I asked to find or do?

- -

(Plan)

How can I use the information I am given?

Which strategy should I try?

- -

(Solve)

Did I follow the plan?

Did I show my work?

Did I write the answer?

- -

(Check)

Did I use the correct information?

Did I do what was asked?

Is my answer reasonable?

Facts   Multiply.

9 ×6	7 ×1	9 ×2	10 ×10	7 ×4	6 ×5	3 ×2	4 ×4	8 ×6	6 ×3
7 ×7	4 ×3	8 ×5	2 ×2	9 ×9	8 ×3	3 ×0	9 ×7	7 ×2	8 ×8
5 ×4	6 ×2	6 ×6	7 ×3	5 ×5	8 ×7	3 ×3	9 ×8	4 ×2	0 ×7
9 ×4	9 ×5	8 ×2	6 ×4	9 ×3	5 ×2	8 ×4	7 ×5	5 ×3	7 ×6

Mental Math

a.	b.	c.	d.
e.	f.	g.	h.

Problem Solving

Understand

What information am I given?
What am I asked to find or do?

Plan

How can I use the information I am given?
Which strategy should I try?

Solve

Did I follow the plan?
Did I show my work?
Did I write the answer?

Check

Did I use the correct information?
Did I do what was asked?
Is my answer reasonable?

Name _____ Time _____

Facts — Divide.

$7\overline{)49}$	$5\overline{)25}$	$3\overline{)27}$	$3\overline{)24}$	$9\overline{)9}$	$3\overline{)12}$	$4\overline{)16}$	$2\overline{)10}$	$6\overline{)42}$	$4\overline{)28}$
$6\overline{)0}$	$2\overline{)4}$	$5\overline{)35}$	$2\overline{)6}$	$3\overline{)15}$	$6\overline{)54}$	$2\overline{)16}$	$8\overline{)72}$	$5\overline{)30}$	$3\overline{)21}$
$3\overline{)9}$	$9\overline{)81}$	$5\overline{)40}$	$4\overline{)20}$	$7\overline{)56}$	$2\overline{)18}$	$6\overline{)36}$	$8\overline{)56}$	$2\overline{)12}$	$7\overline{)42}$
$6\overline{)48}$	$2\overline{)14}$	$4\overline{)36}$	$4\overline{)24}$	$5\overline{)45}$	$2\overline{)8}$	$3\overline{)18}$	$7\overline{)63}$	$4\overline{)32}$	$8\overline{)64}$

Mental Math

a.	b.	c.	d.
e.	f.	g.	h.

Problem Solving

Understand

What information am I given?
What am I asked to find or do?

- -

Plan

How can I use the information I am given?
Which strategy should I try?

- -

Solve

Did I follow the plan?
Did I show my work?
Did I write the answer?

- -

Check

Did I use the correct information?
Did I do what was asked?
Is my answer reasonable?

 Saxon Math Intermediate 5

Facts   Divide.

$9\overline{)81}$	$8\overline{)48}$	$6\overline{)18}$	$8\overline{)40}$	$3\overline{)6}$	$7\overline{)28}$	$5\overline{)15}$	$9\overline{)72}$	$7\overline{)14}$	$5\overline{)25}$
$9\overline{)54}$	$8\overline{)32}$	$4\overline{)12}$	$4\overline{)0}$	$6\overline{)12}$	$4\overline{)16}$	$7\overline{)42}$	$2\overline{)4}$	$9\overline{)45}$	$8\overline{)56}$
$8\overline{)24}$	$9\overline{)63}$	$4\overline{)8}$	$5\overline{)20}$	$3\overline{)9}$	$7\overline{)35}$	$9\overline{)36}$	$8\overline{)16}$	$7\overline{)49}$	$8\overline{)8}$
$6\overline{)42}$	$9\overline{)18}$	$6\overline{)30}$	$7\overline{)21}$	$6\overline{)24}$	$5\overline{)10}$	$6\overline{)36}$	$8\overline{)64}$	$9\overline{)27}$	$7\overline{)56}$

Mental Math

a.	b.	c.	d.
e.	f.	g.	h.

Problem Solving

Understand

What information am I given?

What am I asked to find or do?

Plan

How can I use the information I am given?

Which strategy should I try?

Solve

Did I follow the plan?

Did I show my work?

Did I write the answer?

Check

Did I use the correct information?

Did I do what was asked?

Is my answer reasonable?

Power Up F

Facts   Multiply.

7 × 9	4 × 4	2 × 5	6 × 9	5 × 6	3 × 8	4 × 9	2 × 3	7 × 8	3 × 5
5 × 9	3 × 4	8 × 9	2 × 2	10 × 10	4 × 6	6 × 7	2 × 8	7 × 7	8 × 0
8 × 8	2 × 7	3 × 6	5 × 8	4 × 7	3 × 3	9 × 9	5 × 7	2 × 9	7 × 1
4 × 5	6 × 8	2 × 4	0 × 0	3 × 7	4 × 8	2 × 6	5 × 5	3 × 9	6 × 6

Mental Math

a.	b.	c.	d.
e.	f.	g.	h.

Problem Solving

Understand

What information am I given?

What am I asked to find or do?

- -

Plan

How can I use the information I am given?

Which strategy should I try?

- -

Solve

Did I follow the plan?

Did I show my work?

Did I write the answer?

- -

Check

Did I use the correct information?

Did I do what was asked?

Is my answer reasonable?

       Saxon Math Intermediate 5

Facts . Divide. Write each answer with a remainder.

R 2)7	R 3)16	R 4)15	R 5)28	R 4)21
R 6)15	R 8)20	R 2)15	R 5)43	R 3)20
R 6)27	R 3)25	R 2)17	R 3)10	R 7)30
R 8)25	R 3)8	R 4)30	R 5)32	R 7)50

Mental Math

a.	b.	c.	d.
e.	f.	g.	h.

Problem Solving

Understand

What information am I given?
What am I asked to find or do?

- -

Plan

How can I use the information I am given?
Which strategy should I try?

- -

Solve

Did I follow the plan?
Did I show my work?
Did I write the answer?

- -

Check

Did I use the correct information?
Did I do what was asked?
Is my answer reasonable?

Facts — Write these improper fractions as whole or mixed numbers.

$\frac{8}{3} =$	$\frac{9}{3} =$	$\frac{3}{2} =$	$\frac{4}{3} =$	$\frac{7}{4} =$
$\frac{10}{5} =$	$\frac{10}{9} =$	$\frac{7}{3} =$	$\frac{5}{2} =$	$\frac{11}{8} =$
$\frac{12}{12} =$	$\frac{9}{4} =$	$\frac{12}{5} =$	$\frac{10}{3} =$	$\frac{16}{4} =$
$\frac{13}{5} =$	$\frac{15}{8} =$	$\frac{21}{10} =$	$\frac{9}{2} =$	$\frac{25}{6} =$

Mental Math

a.	b.	c.	d.
e.	f.	g.	h.

Problem Solving

Understand

What information am I given?

What am I asked to find or do?

- -

Plan

How can I use the information I am given?

Which strategy should I try?

- -

Solve

Did I follow the plan?

Did I show my work?

Did I write the answer?

- -

Check

Did I use the correct information?

Did I do what was asked?

Is my answer reasonable?

     Saxon Math Intermediate 5

Name _____ Time _____

Facts   Reduce each fraction to lowest terms.

$\dfrac{2}{10} =$	$\dfrac{3}{9} =$	$\dfrac{2}{4} =$	$\dfrac{6}{8} =$	$\dfrac{4}{12} =$
$\dfrac{6}{9} =$	$\dfrac{4}{8} =$	$\dfrac{2}{6} =$	$\dfrac{3}{6} =$	$\dfrac{6}{10} =$
$\dfrac{5}{10} =$	$\dfrac{3}{12} =$	$\dfrac{2}{8} =$	$\dfrac{4}{6} =$	$\dfrac{50}{100} =$
$\dfrac{2}{12} =$	$\dfrac{8}{16} =$	$\dfrac{9}{12} =$	$\dfrac{25}{100} =$	$\dfrac{6}{12} =$

Mental Math

a.	b.	c.	d.
e.	f.	g.	h.

Problem Solving

Understand

What information am I given?
What am I asked to find or do?

- -

Plan

How can I use the information I am given?
Which strategy should I try?

- -

Solve

Did I follow the plan?
Did I show my work?
Did I write the answer?

- -

Check

Did I use the correct information?
Did I do what was asked?
Is my answer reasonable?

Saxon Math Intermediate 5

Name _____ Time _____

Facts   Simplify.

$\frac{6}{4} =$	$\frac{10}{8} =$	$\frac{9}{12} =$	$\frac{12}{9} =$	$\frac{12}{10} =$
$\frac{12}{8} =$	$\frac{8}{6} =$	$\frac{10}{4} =$	$\frac{8}{20} =$	$\frac{20}{8} =$
$\frac{24}{6} =$	$\frac{9}{6} =$	$\frac{15}{10} =$	$\frac{8}{12} =$	$\frac{10}{6} =$
$\frac{16}{10} =$	$\frac{9}{12} =$	$\frac{15}{6} =$	$\frac{10}{20} =$	$\frac{18}{12} =$

Mental Math

a.	b.	c.	d.
e.	f.	g.	h.

Problem Solving

(Understand)

What information am I given?

What am I asked to find or do?

(Plan)

How can I use the information I am given?

Which strategy should I try?

(Solve)

Did I follow the plan?

Did I show my work?

Did I write the answer?

(Check)

Did I use the correct information?

Did I do what was asked?

Is my answer reasonable?

 Saxon Math Intermediate 5

Facts	Write each percent as a reduced fraction.

50% =	75% =	20% =	5% =
25% =	1% =	60% =	99% =
10% =	90% =	2% =	30% =
40% =	35% =	80% =	4% =

Mental Math

a.	b.	c.	d.
e.	f.	g.	h.

Problem Solving

Understand

What information am I given?
What am I asked to find or do?

Plan

How can I use the information I am given?
Which strategy should I try?

Solve

Did I follow the plan?
Did I show my work?
Did I write the answer?

Check

Did I use the correct information?
Did I do what was asked?
Is my answer reasonable?

Add.

8 + 3	5 + 6	2 + 9	4 + 8	3 + 9	6 + 3	7 + 3	6 + 4
5 + 5	7 + 2	8 + 5	2 + 5	5 + 7	5 + 4	2 + 8	4 + 6
6 + 5	4 + 9	8 + 6	5 + 8	7 + 4	6 + 6	8 + 2	2 + 4
9 + 1	8 + 8	2 + 2	4 + 5	6 + 2	5 + 9	3 + 3	2 + 7
4 + 4	7 + 5	8 + 7	3 + 4	7 + 9	6 + 7	9 + 2	8 + 9
7 + 6	6 + 8	8 + 4	3 + 5	9 + 8	5 + 0	9 + 3	2 + 6
3 + 6	5 + 2	6 + 9	9 + 6	4 + 3	9 + 9	9 + 4	7 + 7
3 + 7	2 + 3	9 + 5	3 + 8	5 + 3	9 + 7	7 + 8	4 + 2

Saxon Math Intermediate 5

Name _____ Time _____

Subtract.

16 − 9	11 − 3	18 − 9	13 − 7	11 − 5	17 − 9	10 − 9	13 − 4
10 − 5	5 − 2	12 − 6	10 − 1	6 − 4	7 − 2	14 − 7	11 − 6
16 − 7	10 − 3	12 − 4	11 − 7	17 − 8	10 − 6	9 − 5	12 − 5
9 − 3	12 − 3	16 − 8	9 − 1	15 − 6	11 − 4	13 − 5	8 − 5
9 − 6	11 − 2	10 − 8	6 − 3	14 − 5	8 − 6	11 − 8	13 − 8
7 − 4	10 − 7	12 − 8	8 − 7	7 − 3	7 − 6	7 − 5	8 − 4
13 − 6	15 − 8	13 − 9	11 − 9	10 − 4	9 − 4	10 − 2	8 − 3
14 − 8	12 − 9	9 − 8	12 − 7	15 − 7	14 − 9	9 − 7	8 − 2

Name _____ Time _____

Multiply.

9 × 9	3 × 5	8 × 5	2 × 6	4 × 7	7 × 2	7 × 8	3 × 4
5 × 9	7 × 3	2 × 7	6 × 3	5 × 4	1 × 0	9 × 2	9 × 0
2 × 8	6 × 4	8 × 1	3 × 3	4 × 8	9 × 3	4 × 9	8 × 4
6 × 5	2 × 9	9 × 4	7 × 4	5 × 8	4 × 2	9 × 8	3 × 6
5 × 5	5 × 0	6 × 6	7 × 9	9 × 1	5 × 1	4 × 3	8 × 9
3 × 7	9 × 7	7 × 5	8 × 8	8 × 3	5 × 2	9 × 5	6 × 7
8 × 6	3 × 8	7 × 6	9 × 6	4 × 4	5 × 3	7 × 7	6 × 2
4 × 5	6 × 8	8 × 7	4 × 6	5 × 7	8 × 2	6 × 9	3 × 9

28

Divide.

7)21	2)10	6)42	1)3	4)24	9)54	6)18	5)30
4)32	8)56	6)12	3)18	9)72	5)15	7)42	6)36
5)10	2)6	7)63	4)16	8)48	1)2	5)35	3)21
2)18	3)15	8)40	5)20	9)27	7)35	4)20	9)63
1)4	7)14	8)24	6)24	2)16	8)64	4)28	7)49
2)4	9)81	3)12	6)30	8)32	9)36	3)27	2)14
5)25	6)48	7)28	4)36	5)45	4)8	8)16	3)24
9)45	6)54	7)56	8)72	5)40	3)9	9)18	4)12

Selecting Tools and Techniques

Time Requirement
10–15 minutes

Materials/Preparation
- Test-Day Activity Student Worksheet 1 (1 per student)

Activity

Explain to students that they will be buying supplies at a stationary store. They will choose mental math, paper and pencil, or calculator to solve a problem and be asked to explain their choice. Explain that all of the information needed is on **Test-Day Activity 1**.

Selecting Tools and Techniques

You are at a stationary store buying supplies. Read each problem below and state if you would use mental math, paper and pencil, or calculator to solve the problem. Explain why. Use each method once.

1. You need to buy a package of pens for $6 and a package of paper for $2. You want to find out how much these two items will cost together.

   _____

   _____

2. You have selected the following items to buy:

   package of pencils $3.98

   package of yellow paper. $2.13

   package of envelopes. $3.48

   package of paper clips $0.35

   sticky notes. $0.74

   calendar $9.29

   You want to find out exactly how much the total will be before you pay for them.

   _____

   _____

3. You need to buy an item for $2.78 and another item for $1.92. You want to find the total cost of the two items. You discover that your calculator is broken.

   _____

   _____

Dividing Fractions

Time Requirement

10–15 minutes

Materials/Preparation

• Test-Day Activity Student Worksheet 2 (1 per student)

Activity

Explain to students that they will use an area model to divide fractions. They will be asked to explain how they used a model to solve a problem. If necessary, review how to divide fractions before doing this activity. Point out that all of the information needed is on **Test-Day Activity 2**.

Dividing Fractions

A question we can ask when we divide is, "How many . . . . make . . . .?"

For example, $12 \div 3$ can be read, "How many 3s make 12?"

Using a model we can count how many 3s make 12.

We can see that 4 groups of 3 make 12.

We can ask a similar question when we divide some fractions.

We can read $\frac{1}{2} \div \frac{1}{4}$ as, "How many $\frac{1}{4}$s make $\frac{1}{2}$?"

Using pictures or fraction manipulatives we can model the problem.

$$\frac{1}{2} \div \frac{1}{4} = ?$$

We see that two one-fourths make half of a circle.

$$\frac{1}{2} \div \frac{1}{4} = 2$$

Use a model to divide.

1. $2 \div \frac{1}{2} = ?$

Divide the two circles below to show how many $\frac{1}{2}$s make 2.

2. Explain how you found the answer to problem **1.**

_____

3. $\frac{3}{4} \div \frac{1}{8} = ?$

How many $\frac{1}{8}$s make $\frac{3}{4}$? One eighth of a circle is . One eighth is half of $\frac{1}{4}$. Use the drawing of $\frac{3}{4}$ of a circle below to make a model of the problem.

_____

 Saxon Math Intermediate 5

Congruent Geometric Figures

Time Requirement

10–15 minutes

Materials/Preparation

- Test-Day Activity Student Worksheet 3 (1 per student)

Activity

Explain to students that they will identify congruent figures, and corresponding sides, vertices, and angles. They will be asked to explain how they know two figures are congruent. If necessary, review corresponding sides, vertices, and angles in congruent figures before the doing this activity. Explain that all of the information needed is on **Test-Day Activity 3**.

Name _____

Date _____

Congruent Geometric Figures

These two triangles are congruent because they are the same size and shape.

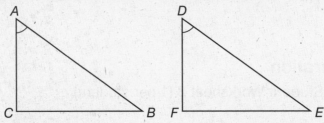

We can name these triangles using the letters at each vertex. We say that triangle *ABC* is congruent to triangle *DEF*.

We also say that angle *A* corresponds to angle *D* and that side $\overline{AB}$ corresponds to side $\overline{DE}$.

When we name congruent polygons and their parts with letters, we are careful to name the vertices in the same order.

Look at the triangles below to answer problems **1** and **2**.

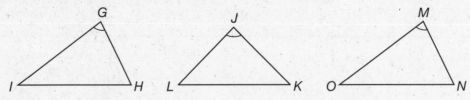

1. Triangle *GHI* is congruent to which other triangle? Explain.

   _____

2. In the congruent triangles,

   a. vertex *N* corresponds to _____

   b. side $\overline{HI}$ corresponds to _____

   c. angle *M* corresponds to _____

3. Name one pair of corresponding angles, and one pair of corresponding sides in the congruent polygons shown at right.

    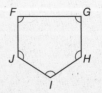

   _____

Saxon Math Intermediate 5

Multiplying Fractions

Time Requirement

10–15 minutes

Materials/Preparation

- Test-Day Activity Student Worksheet 4 (1 per student)
- Grid paper

Activity

Explain to students that they will use an area model to multiply two fractions. They will be asked to explain how they used a model to solve a problem. Demonstrate how to multiply two fractions on a grid before releasing students on this activity. Some students may benefit from using grid paper. Point out that all of the information needed is on **Test-Day Activity 4**.

Name _____

Date _____

Multiplying Fractions

We can use a square model to multiply two fractions. First we draw a square to represent 1×1. Then we divide two sides of the square to represent the two fractions we are multiplying.

Here is how we model $\frac{2}{3} \times \frac{3}{4}$. We divide one side of the square vertically into 3 equal parts and divide the other side of the square horizontally into 4 equal parts.

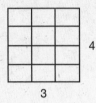

Now we mark off $\frac{2}{3}$ and $\frac{3}{4}$. As we darken the lines at $\frac{2}{3}$ and $\frac{3}{4}$, we see that they make up 6 of the 12 parts of the square.

$$\frac{2}{3} \times \frac{3}{4} = \frac{6}{12} = \frac{1}{2}$$

This means that $\frac{2}{3}$ of $\frac{3}{4}$ is $\frac{6}{12}$. Another name for $\frac{6}{12}$ is $\frac{1}{2}$.

For problems **1** and **2**, divide the square to make a model of the problem. Then write the multiplication on the line.

1. $\frac{1}{2} \times \frac{1}{4} = ?$

_____

2. $\frac{1}{2} \times \frac{3}{4} = ?$

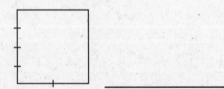

_____

3. The multiplication $\frac{1}{2} \times \frac{1}{4}$ means "$\frac{1}{2}$ of $\frac{1}{4}$". How could you use your fractions manipulatives to find $\frac{1}{2}$ of $\frac{1}{4}$?

Saxon Math Intermediate 5

Finding Common Denominators

Time Requirement

10–15 minutes

Materials/Preparation

• Test-Day Activity Student Worksheet 5 (1 per student)

Activity

Explain to students that they will compare fractions. They will use their fraction manipulatives to help them find common denominators. They will also use picture models. If necessary, review how to find fractions with common denominators before doing this activity. Point out that all of the information needed is on **Test-Day Activity 5**.

Finding Common Denominators

The same fraction can be named different ways. Below are three ways to name $\frac{1}{3}$. Renaming a fraction changes the numerator and denominator, but not the size of the fraction.

By renaming fractions we can find a way to name two fractions so that they have the same denominator.

Use your fraction manipulatives to help solve the following problems.

1. The fractions $\frac{1}{2}$ and $\frac{3}{4}$ do not have common denominators. Rename $\frac{1}{2}$ so that its denominator is 4. _____

2. The fractions $\frac{1}{3}$ and $\frac{5}{12}$ do not have common denominators. Find how many twelfths equal $\frac{1}{3}$. _____

3. The fractions $\frac{5}{8}$ and $\frac{3}{4}$ do not have common denominators. Find a fractions equal to $\frac{3}{4}$ that has 8 as its denominator. _____

Use the pictures to find fractions with common denominators for each pair of fractions. Compare the fractions.

4. Rename $\frac{1}{2}$ and $\frac{2}{3}$ so that both fractions have a denominator of 6. _____

5. Write $\frac{2}{3}$ and $\frac{5}{6}$ with common denominators. _____

6. Write $\frac{1}{2}$ and $\frac{1}{6}$ with common denominators. _____

7. Rename $\frac{1}{2}$, $\frac{1}{3}$, and $\frac{1}{4}$ so that all three fractions have a common denominator of 12. _____

 Saxon Math Intermediate 5

Making and Comparing Graphs

Time Requirement

10–15 minutes

Materials/Preparation

- Test-Day Activity Student Worksheet 6 (1 per student)

Activity

Explain to students that they will be making graphs to show the number of different kinds of toy robots sold at a toy store in one day. They will make a bar graph and a circle graph to show the data and will discuss the advantages and disadvantages of each type of graph. Point out that all of the information needed is on **Test-Day Activity 6**.

Name _____

Date _____

Making and Comparing Graphs

The following table shows how many people bought each kind of toy robot one day at a toy store.

Toy Robot Sales

Kind of Robot	Number Sold
Dog Robot	8
Dinosaur Robot	16
Person Robot	4
Car Robot	4

1. Make a bar graph to show the robot sales information in the table.

Toy Robots Sold

Kind of Toy Robot

2. Altogether 32 robot toys were sold during the day. Complete the circle graph to show the fraction of the total sales each kind of robot represents.

3. Compare the bar graph to the circle graph. What are the advantages and disadvantages of each?

_____

_____

Saxon Math Intermediate 5

Collecting, Organizing, and Displaying Data

Time Requirement

10–15 minutes

Materials/Preparation

- Test-Day Activity Student Worksheet 7 (1 per student)

Task

Explain to students that they will organize and display data in an appropriate graph. They will be asked to explain their selection of the kind graph to display the data. Point out that all of the information needed is on **Test-Day Activity 7**.

Collecting, Organizing, and Displaying Data

The table shows the results of a survey in which people were asked which city they would most like to visit.

The City People Want to Visit Most

Name of City	Number of People
Paris	19
New York	15
Tokyo	24
Mexico City	21

1. Decide which kind of graph you would like to use to display this data. Make the graph.

2. Explain why you chose the graph in problem **1**.

_____

_____

_____

       Saxon Math Intermediate 5

Area of a Regular Polygon

Time Requirement

10–15 minutes

Materials/Preparation

- Test-Day Activity Student Worksheet 8 (1 per student)

Activity

Explain to students that they will use their knowledge of area formulas for squares, rectangles, and triangles to find the area of a polygon with these shapes. They will be asked to explain how to use these formulas to find the area of a regular polygon. If necessary, review the formulas for the area of a square, rectangle, and triangle before the students do this activity. Explain that all of the information needed is on **Test-Day Activity 8**.

Name _____

Date _____

Area of a Regular Polygon

These are the formulas for the area of squares, rectangles, and triangles.

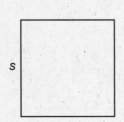 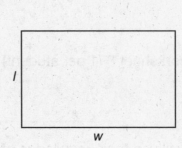 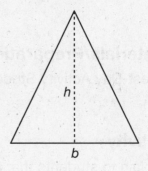

$A = s^2$ $A = l \times w$ $A = \frac{1}{2} \times (b \times h)$

1. Show how to divide this polygon into two parts. Use the formula for the area of a rectangle to find the area of each part. Then add the areas of each part to find the area of the polygon.

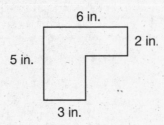

_____

_____

2. Find the area of the polygon below. Describe how you found the area.

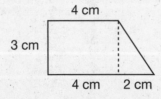

_____

_____

           Saxon Math Intermediate 5

Area of a Circle

Time Requirement

10–15 minutes

Materials/Preparation

- Test-Day Activity Student Worksheet 9 (1 per student)

Activity

Explain to students that they will estimate the area of a circle that is drawn on a grid. They will be asked to explain how they estimated the area of a circle on a grid and how square units can be used to measure the area of a figure that is round. Point out that all of the information needed is on **Test-Day Activity 9**. You might want to allow the students to use calculators for the computation.

Name _____

Date _____

Area of a Circle

You can use a grid to estimate the area of a circle. Each square in the grid represents one square centimeter.

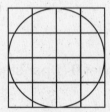

There are about 12 whole square centimeters in the circle. The circle also covers parts of four other squares. Together these parts equal about one more square centimeter. So the area of this circle is about 13 square centimeters.

1. Use the grid to estimate the area of the circle. Explain your answer.

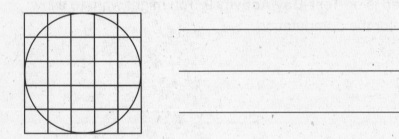

_____

_____

2. Use the grid to estimate the area of the swimming pool in Jamie's backyard.

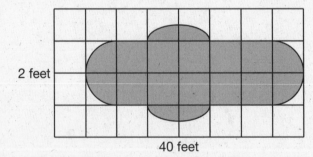

2 feet

40 feet

_____ square feet

Explain how using a square grid can help you determine the area of irregular shapes.

_____

_____

Saxon Math Intermediate 5

Cubic Units

Time Requirement
10–15 minutes

Materials/Preparation
- Test-Day Activity Student Worksheet 10 (1 per student)

Activity
Explain to students that they will name cubic units that are pictured. They will also state the appropriate unit to measure length, area, and volume and to explain why they chose that unit. If necessary, review cubic units and how they are used to measure volume. Explain that all of the information needed is on **Test-Day Activity 10**.

Name _____

Date _____

Cubic Units

The amount of space an object occupies is called its **volume**. **Cubic units** are used to measure volume. The figure to the right illustrates a cubic inch (in.³) because it is a cube where all three dimensions measure 1 inch.

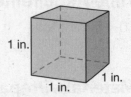

Name the cubic unit pictured in each problem below.

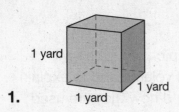

1. _____

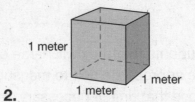

2. _____

We use different kinds of units to measure lengths, area, and volumes.

1 centimeter	1 square centimeter	1 cubic centimeter

measures length   measures area   measures volume

3. Which unit would you use to measure the length of a pencil? Choose a unit and explain your choice.

    A. centimeter   **B.** square centimeter   **C.** cubic centimeter

_____

4. Which unit would you use to measure the area of this paper? Choose a unit and explain your choice.

    A. centimeter   **B.** square centimeter   **C.** cubic centimeter

_____

 Saxon Math Intermediate 5

Formulas

Time Requirement
10–15 minutes

Materials/Preparation
- Test-Day Activity Student Worksheet 11 (1 per student)

Activity
Explain to students that they will find the area of two rectangles by using pictures and the formula for the area of a rectangle. They will draw a conclusion about whether the formula for the area of a rectangle is reliable regardless of what type of number is used. Students will substitute different numbers for the unknowns in various formulas. If necessary, review, area of rectangles, multiplying decimals, and/or finding the value of expressions, before doing this activity. Explain that all of the information needed is on **Test-Day Activity 11**.

Name _____

Date _____

Formulas

Recall that a formula is a mathematical description of how to find a measure or number. We put numbers we know into a formula to find the number we want to know.

1. Use the formula for the area of a rectangle $A = l \times w$, to find the area (A) when the length (l) is 5 ft and the width (w) is 3 ft.

   _____

2. Use the formula for the area of a triangle, $A = \frac{1}{2}bh$ to find the area (A) of a triangle with a base (b) of 8 inches and a height (h) of 6 inches.

   _____

A formula works for whole numbers, for fractions, and for decimal numbers.

3. The two rectangles below are congruent, but their length and width are named with different units. Use the area formula $V = l \times w$ to find the area of each rectangle.

   3 mm ▢ 4 mm   0.3 cm ▢ 0.4 cm

   _____   _____

4. Your two answers in problem **3** should be equal. Here we show an enlarged view of a square centimeter. It is also 100 square millimeters. Draw a 4 mm long and 3 mm wide rectangle on the square. Explain why your two answers to problem **3** are equal.

Saxon Math Intermediate 5

Name _____

For use with Lessons 3, 4, 5, 6, 7, 9, 10, and 26

One-Dollar Bills

Name _____

For use with Lessons 3, 4, 5, 6, 7, 9, 10, and 26

Ten-Dollar Bills

Saxon Math Intermediate 5

Name _____

One Hundred-Dollar Bills

Currency

Saxon Math Intermediate 5

Using Money to Model Arithmetic

Quarters

quarter $\frac{1}{4}$	quarter $\frac{1}{4}$	quarter $\frac{1}{4}$
quarter $\frac{1}{4}$	quarter $\frac{1}{4}$	quarter $\frac{1}{4}$
quarter $\frac{1}{4}$	quarter $\frac{1}{4}$	quarter $\frac{1}{4}$
quarter $\frac{1}{4}$	quarter $\frac{1}{4}$	quarter $\frac{1}{4}$
quarter $\frac{1}{4}$	quarter $\frac{1}{4}$	quarter $\frac{1}{4}$
quarter $\frac{1}{4}$	quarter $\frac{1}{4}$	quarter $\frac{1}{4}$

Dimes and Pennies

dime		$\frac{1}{10}$	dime		$\frac{1}{10}$	penny		$\frac{1}{100}$
dime		$\frac{1}{10}$	dime		$\frac{1}{10}$	penny		$\frac{1}{100}$
dime		$\frac{1}{10}$	dime		$\frac{1}{10}$	penny		$\frac{1}{100}$
dime		$\frac{1}{10}$	dime		$\frac{1}{10}$	penny		$\frac{1}{100}$
dime		$\frac{1}{10}$	dime		$\frac{1}{10}$	penny		$\frac{1}{100}$
dime		$\frac{1}{10}$	dime		$\frac{1}{10}$	penny		$\frac{1}{100}$
dime		$\frac{1}{10}$	dime		$\frac{1}{10}$	penny		$\frac{1}{100}$
dime		$\frac{1}{10}$	dime		$\frac{1}{10}$	penny		$\frac{1}{100}$
dime		$\frac{1}{10}$	dime		$\frac{1}{10}$	penny		$\frac{1}{100}$
dime		$\frac{1}{10}$	dime		$\frac{1}{10}$	penny		$\frac{1}{100}$

Place-Value Template

Place-Value Template

| hundreds | tens | ones |

Name _____

Place Value Charts

hundred thousands

ten thousands

thousands

hundreds

tens

ones

___ ___ ___ , ___ ___ ___

hundred billions

ten billions

billions

hundred millions

ten millions

millions

hundred thousands

ten thousands

thousands

hundreds

tens

ones

___ ___ ___ , ___ ___ ___ , ___ ___ ___ , ___ ___ ___

Billions				Millions				Thousands				Units (Ones)		
hundreds	tens	ones	billions comma	hundreds	tens	ones	millions comma	hundreds	tens	ones	thousands comma	hundreds	tens	ones
___	___	___		___	___	___		___	___	___		___	___	___

Checks

_____

My Street
My Town, USA

0001

DATE _____

PAY TO THE
ORDER OF _____ $ [_____]

_____ DOLLARS

***MY**BANK

Memo_____   _____

THIS CHECK IS NONNEGOTIABLE, FOR EDUCATION PURPOSES ONLY

_____

My Street
My Town, USA

0002

DATE _____

PAY TO THE
ORDER OF _____ $ [_____]

_____ DOLLARS

***MY**BANK

Memo_____   _____

THIS CHECK IS NONNEGOTIABLE, FOR EDUCATION PURPOSES ONLY

_____

My Street
My Town, USA

0003

DATE _____

PAY TO THE
ORDER OF _____ $ [_____]

_____ DOLLARS

***MY**BANK

Memo_____   _____

THIS CHECK IS NONNEGOTIABLE, FOR EDUCATION PURPOSES ONLY

Number Lines

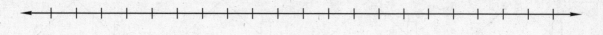

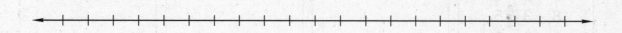

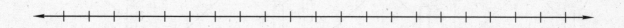

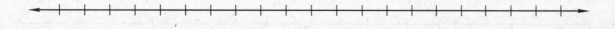

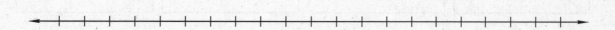

Fact Family Houses

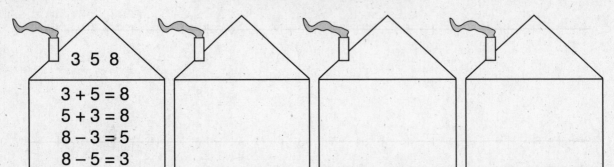

3 5 8

$3 + 5 = 8$
$5 + 3 = 8$
$8 - 3 = 5$
$8 - 5 = 3$

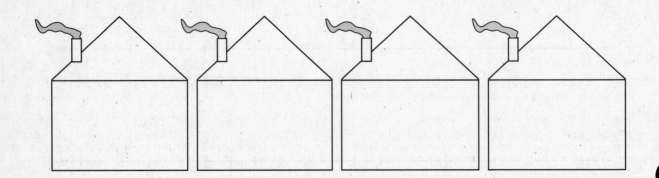

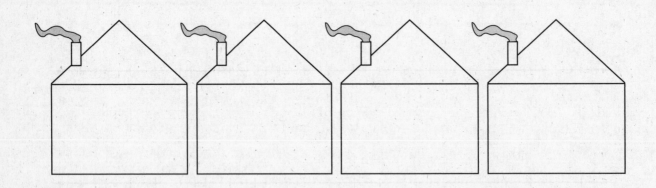

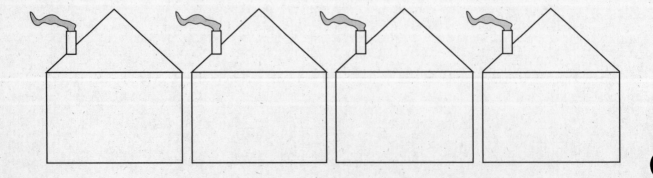

Saxon Math Intermediate 5

Problems About Combining

a. The troop hiked 8 miles in the morning.

b. The troop hiked 7 miles in the afternoon.

c. Altogether, the troop hiked 15 miles in the morning and afternoon.

Complete the question that replaces the missing statement in each of the following paragraphs.

1. Sentence **a** is missing:

The troop hiked 7 miles in the afternoon. Altogether, the troop hiked 15 miles in the morning and afternoon.

How many miles _____

_____

_____

2. Sentence **b** is missing:

The troop hiked 8 miles in the morning. Altogether, the troop hiked 15 miles in the morning and afternoon.

How many miles _____

_____

_____

3. Sentence **c** is missing:

The troop hiked 8 miles in the morning. The troop hiked 7 miles in the afternoon.

Altogether, how many miles _____

_____

_____

Problems About Separating

d. Jack went to the store with $28.

e. Jack spent $12 at the store.

f. Jack left the store with $16.

Complete the question that replaces the missing statement in each of the following paragraphs.

4. Sentence **d** is missing:

Jack spent $12 at the store. Jack left the store with $16.

How much money _____

_____

_____

5. Sentence **e** is missing:

Jack went to the store with $28. Jack left the store with $16.

How much money _____

_____

_____

6. Sentence **f** is missing:

Jack went to the store with $28. Jack spent $12 at the store.

How much money _____

_____

_____

Problems About Equal Groups

g. At Lincoln School there are 4 classes of fifth grade students.

h. There are 30 students in each fifth grade class.

i. Altogether, there are 120 fifth grade students at Lincoln School.

Complete the question that replaces the missing statement in each of the following paragraphs.

7. Sentence **g** is missing:

There are 30 students in each fifth grade class. Altogether, there are 120 fifth grade students at Lincoln School.

How many classes _____

_____

_____

8. Sentence **h** is missing:

At Lincoln School there are 4 classes of fifth grade students. Altogether, there are 120 fifth grade students at Lincoln School.

How many classes _____

_____

_____

9. Sentence **i** is missing:

At Lincoln School there are 4 classes of fifth grade students. There are 30 students in each fifth grade class.

Altogether, how many _____

_____

_____

Problems About Comparing

j. Abe is 5 years old.

k. Gabe is 11 years old.

l. Version 1: Gabe is 6 years older than Abe.

l. Version 2: Abe is 6 years younger than Gabe.

Complete the question that replaces the missing statement in each of the following paragraphs.

10. Sentence **j** is missing:

   Gabe is 11 years old. Gabe is 6 years older than Abe.

   How old _____

   _____

11. Sentence **k** is missing:

   Abe is 5 years old. Abe is 6 years younger than Gabe.

   How old _____

   _____

12. Sentence **l**, version 1 is missing:

   Abe is 5 years old. Gabe is 11 years old.

   Gabe is _____

   _____

13. Sentence **l**, version 2 is missing:

   Abe is 5 years old. Gabe is 11 years old.

   Abe is _____

   _____

Name _____

Word Problems

1. Below is a three-frame story about Arnold's trip to the store. Can you help Arnold

find out how much money he will get back? _____

2. Write a combining word problem that can be solved by adding.

_____

_____

3. Write a separating word problem that can be solved by subtracting.

_____

_____

4. Write an "equal groups" word problem that can be solved by multiplying.

_____

_____

5. Write an "equal groups" word problem that can be solved by dividing.

_____

_____

6. Write a comparison word problem that can be solved by subtracting.

_____

_____

7. Select one written word problem from problems **2-6** and illustrate it in three
frames.

Celsius and Fahrenheit Thermometers

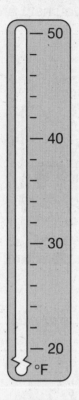

Saxon Math Intermediate 5

Name _____

Clock Face

Name _____

Ruler

Saxon Math Intermediate 5

Hundred Number Chart

1	2	3	4	5	6	7	8	9	10
11	12	13	14	15	16	17	18	19	20
21	22	23	24	25	26	27	28	29	30
31	32	33	34	35	36	37	38	39	40
41	42	43	44	45	46	47	48	49	50
51	52	53	54	55	56	57	58	59	60
61	62	63	64	65	66	67	68	69	70
71	72	73	74	75	76	77	78	79	80
81	82	83	84	85	86	87	88	89	90
91	92	93	94	95	96	97	98	99	100

Name _____

Making a Multiplication Table

	0	1	2	3	4	5	6	7	8	9	10
0											
1											
2											
3											
4											
5											
6											
7											
8											
9											
10											

Saxon Math Intermediate 5

Multiplication Table

	0	1	2	3	4	5	6	7	8	9	10	11	12
0	0	0	0	0	0	0	0	0	0	0	0	0	0
1	0	1	2	3	4	5	6	7	8	9	10	11	12
2	0	2	4	6	8	10	12	14	16	18	20	22	24
3	0	3	6	9	12	15	18	21	24	27	30	33	36
4	0	4	8	12	16	20	24	28	32	36	40	44	48
5	0	5	10	15	20	25	30	35	40	45	50	55	60
6	0	6	12	18	24	30	36	42	48	54	60	66	72
7	0	7	14	21	28	35	42	49	56	63	70	77	84
8	0	8	16	24	32	40	48	56	64	72	80	88	96
9	0	9	18	27	36	45	54	63	72	81	90	99	108
10	0	10	20	30	40	50	60	70	80	90	100	110	120
11	0	11	22	33	44	55	66	77	88	99	110	121	132
12	0	12	24	36	48	60	72	84	96	108	120	132	144

Name _____

Halves

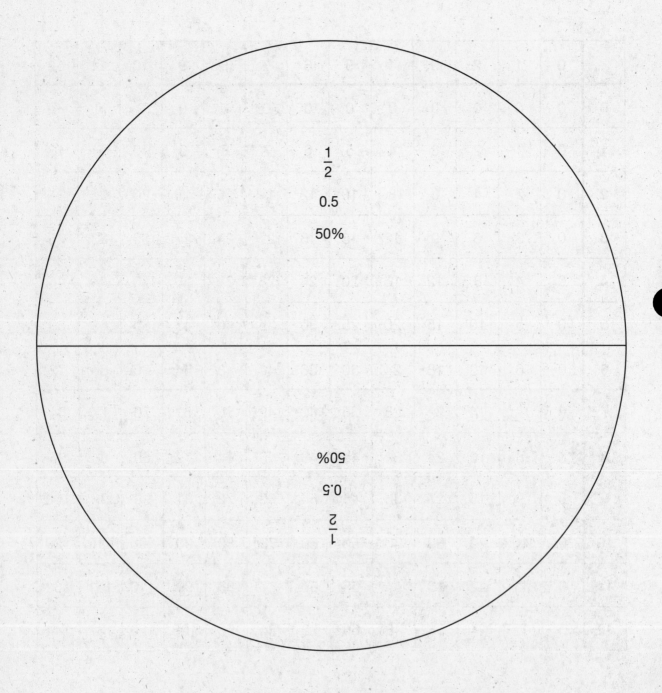

$\frac{1}{2}$

0.5

50%

50%

0.5

$\frac{1}{2}$

Name _____

For use with Investigations 2 and 3, and
Lessons 30, 41, 59, 71, 76, 81, and 87

Fourths

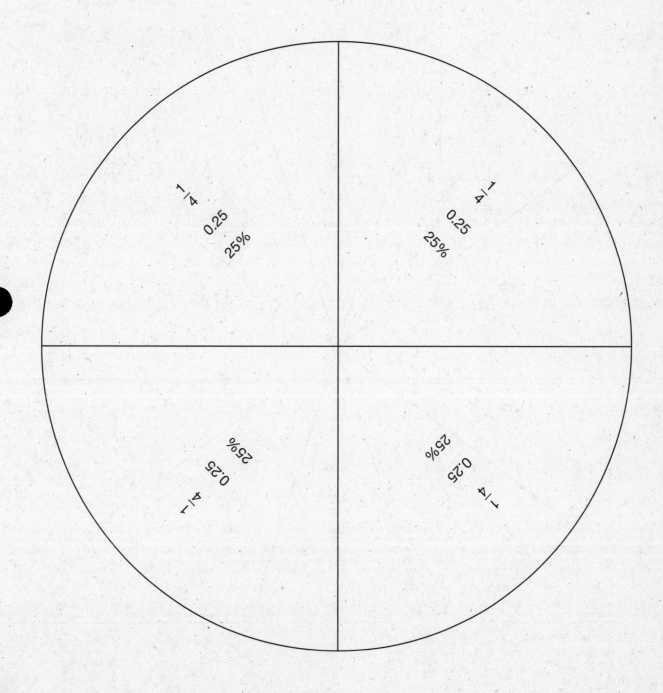

Name _____

Tenths

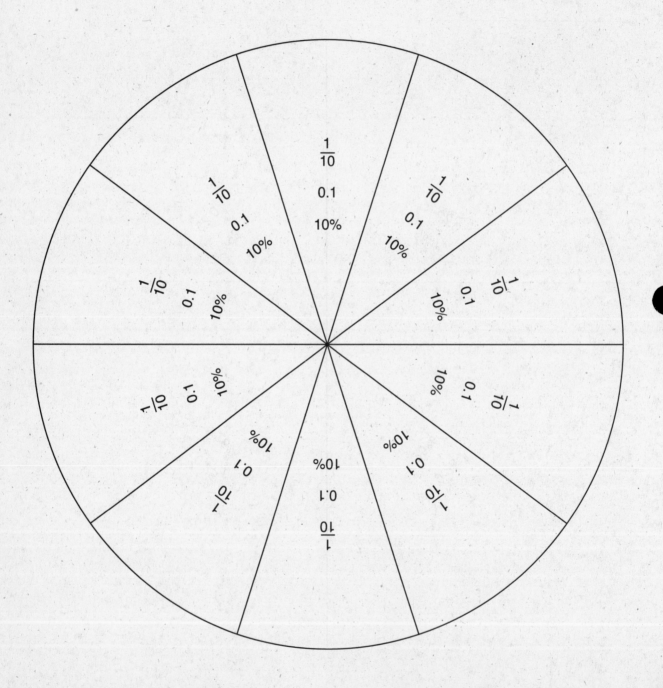

Saxon Math Intermediate 5

Name _____

Thirds

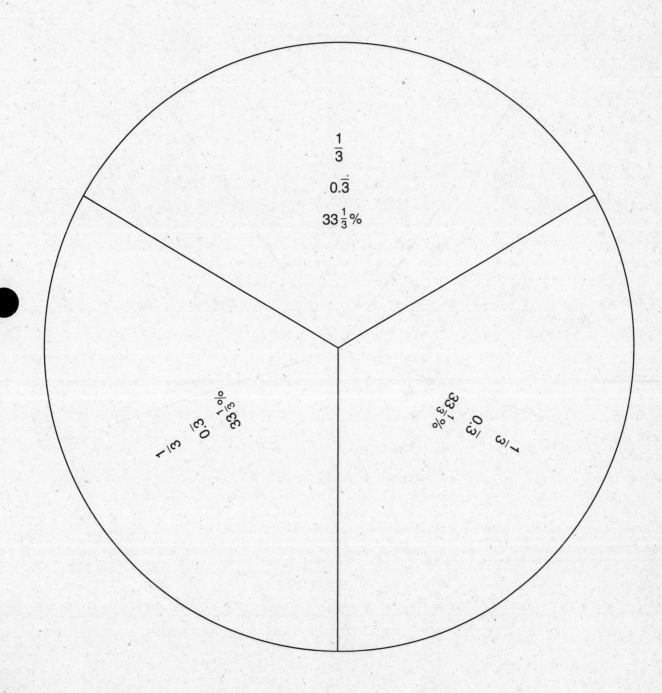

$\dfrac{1}{3}$

$0.\overline{3}$

$33\dfrac{1}{3}\%$

Name _____

Fifths

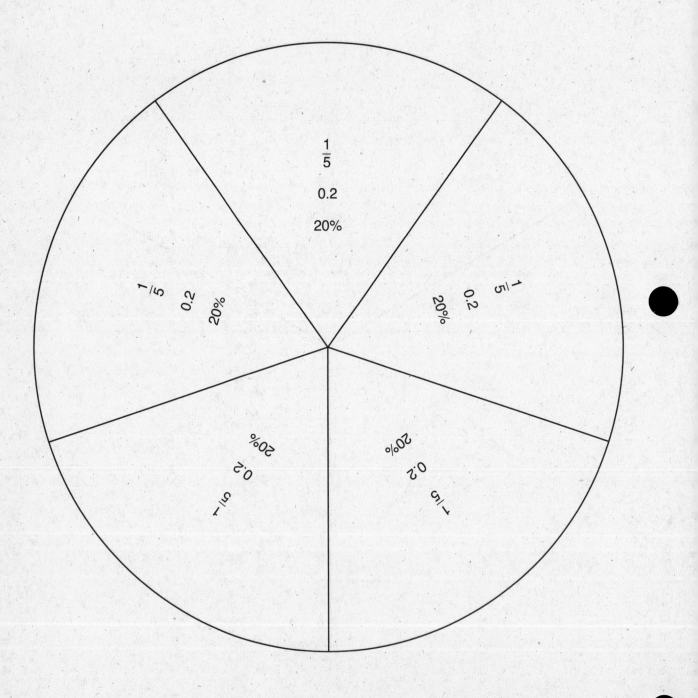

 Saxon Math Intermediate 5

Eighths

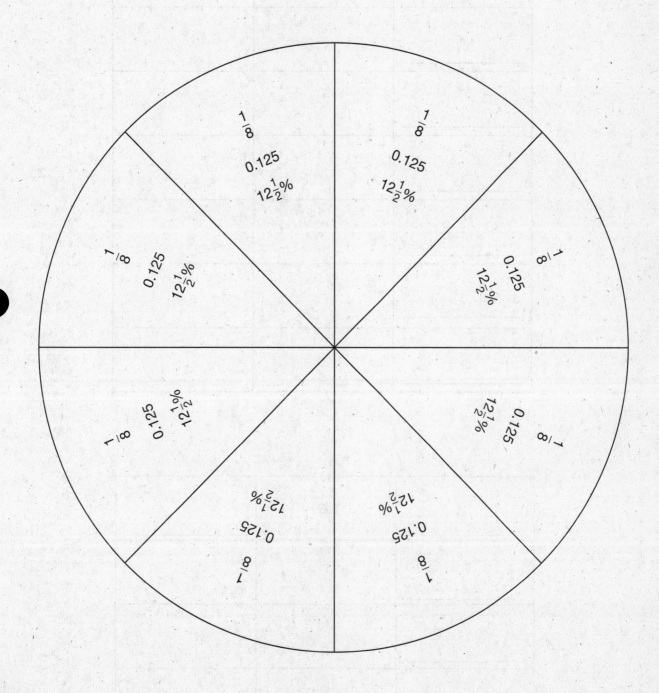

Frequency Tables

_____	Tally	Frequency
_____		
_____		
_____		
_____		
_____		
_____		

_____	Tally	_____
_____		
_____		
_____		

Name _____

Line Graphs

(Title) _____

( )
( )

(Label)

(Label) _____ ( )

(Title) _____

( )
( )

(Label)

(Label) _____ ( )

Pictographs and Bar Graphs

(Title)	

Key: _____

(Title) _____

(Label)

(Label) _____

Horizontal Bar Graph

Give your graph a title.
Label your categories.
Label the horizontal axis with your scale.
Draw more lines on the graph if necessary to show the correct increments.

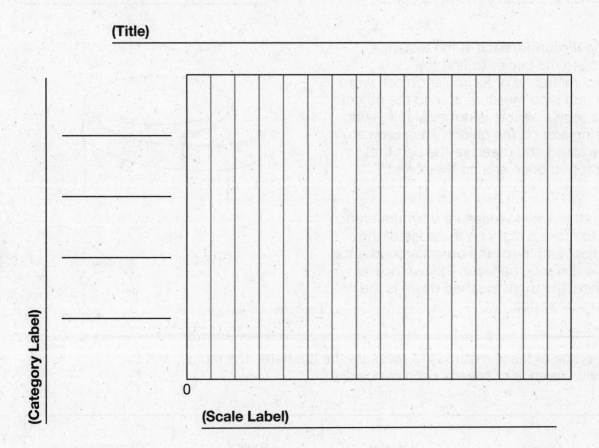

(Title) _____

(Category Label)

0

(Scale Label) _____

Measuring Circles

To measure the radius and the diameter of a circle or circular object, first locate the center of the circle. The **radius** is the distance from the center to the circle. The **diameter** is the distance across the circle through its center.

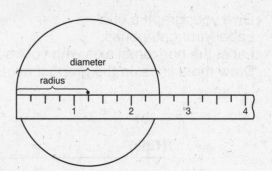

The **circumference** is the distance around the circle. To find the circumference of a circular object, wrap a cloth tape measure around the object. If a tape measure is not available, wrap string around the object. Then unwrap the string and measure the part that wrapped once around the object.

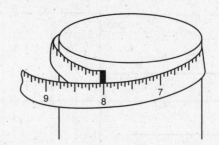

Another way to measure a circumference is to make a mark on the edge of the object and to roll the object around once. The distance between the two points where the mark touched down is the circumference.

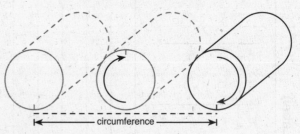

Use one of these methods to measure the diameter, the radius, and the circumference of several circular objects.

Object	Diameter	Radius	Circumference

Saxon Math Intermediate 5

Metric Unit Strips

Below are decimeter, centimeter, and millimeter strips. Cut out the strips and glue or tape them to Lesson Activity Master 22 to make the five lengths shown. You will need to cut the centimeter strips and millimeter strips into smaller pieces to complete the activity.

1 decimeter (10 cm, 100 mm)
1 decimeter (10 cm, 100 mm)
1 decimeter (10 cm, 100 mm)
1 decimeter (10 cm, 100 mm)
1 decimeter (10 cm, 100 mm)

1 cm	1 cm	1 cm	1 cm	1 cm	1 cm	1 cm	1 cm	1 cm	1 cm

1 cm	1 cm	1 cm	1 cm	1 cm	1 cm	1 cm	1 cm	1 cm	1 cm

1 cm	1 cm	1 cm	1 cm	1 cm	1 cm	1 cm	1 cm	1 cm	1 cm

10 mm	10 mm	10 mm	10 mm	10 mm	10 mm	10 mm	10 mm	10 mm	10 mm

Decimal Parts of a Meter

Paste or tape the metric unit strips from Lesson Activity 35 onto this page to make each length. The first length is marked as an example to show which strips to use and where to paste them. Before making each length, decide which metric units to use and how many of each unit are needed. You will need to cut the centimeter and millimeter strips into smaller pieces to complete this activity.

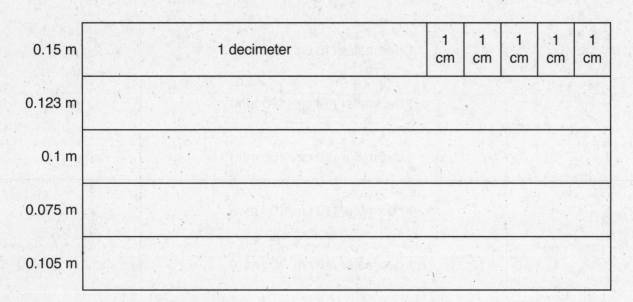

Refer to the lengths you made to complete these problems:

1. Write the five lengths above in order from shortest to longest.

   _____ , _____ , _____ , _____ , _____

2. Compare:

   a. 0.15 m ◯ 0.123 m   **b.** 0.075 m ◯ 0.1 m

   c. 0.15 m ◯ 0.075 m   **d.** 0.150 m ◯ 0.15 m

3. Convert:

   a. 0.15 meters is how many centimeters? _____

   b. 0.123 meters is how many millimeters? _____

4. Round:

   a. Is 0.123 meters closer in length to 0.1 meters or 0.15 meters? _____

   b. Is 0.123 meters closer in length to 0.1 meters or 0.2 meters? _____

5. Add by counting the parts:

   a. 0.15 m + 0.123 m = _____   **b.** 0.15 m + 0.1 m = _____

Saxon Math Intermediate 5

Name _____

Base Ten Squares

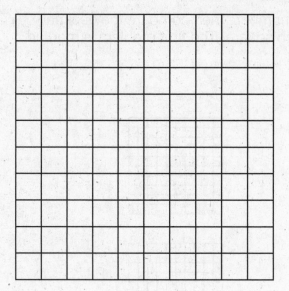
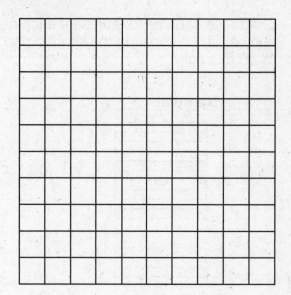

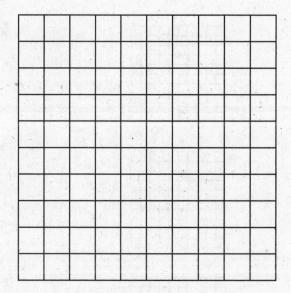
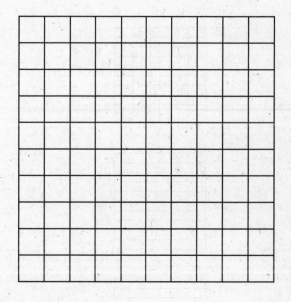

Saxon Math Intermediate 5

Name _____

Comparing Decimal Numbers

Shade the squares to represent each decimal number. Then compare the decimal numbers by comparing the shaded part of each square. The first problem is marked as an example.

Example:

   0.4 $>$ 0.33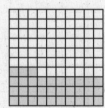

1.

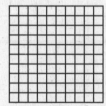 ... 0.1 $\bigcirc$ 0.01

2.   0.12 $\bigcirc$ 0.21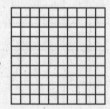

3.    0.3 $\bigcirc$ 0.30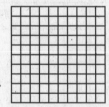

4.    0.5 $\bigcirc$ 0.05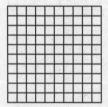

5.    0.6 $\bigcirc$ 0.67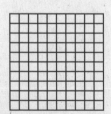

Saxon Math Intermediate 5

Name _____

1-Inch Grid

1-Centimeter Grid

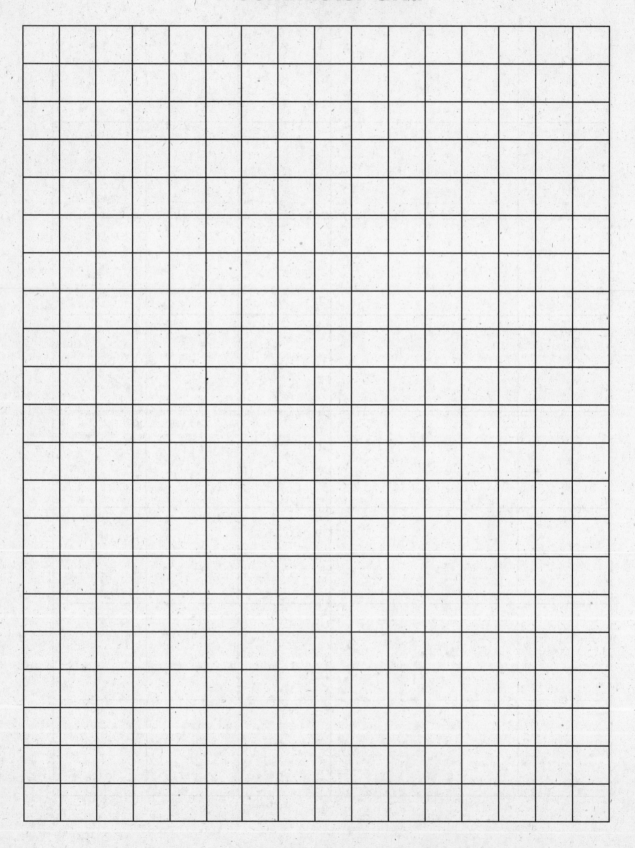

Name _____

Coordinate Plane

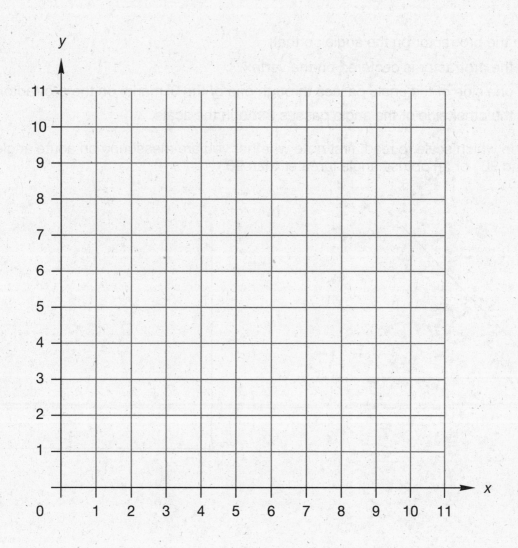

Measuring Angles

Use a protractor to measure each angle on this page. Write your measurements inside each angle.

Position the protractor on the angle so that:

1. the protractor is centered on the vertex.

2. one side of the angle passes through one of the 0° marks on the protractor.

3. the other side of the angle passes through the scale.

To decide which scale to read, first note whether you are measuring an acute angle (less than 90°) or an obtuse angle (greater than 90°).

a.

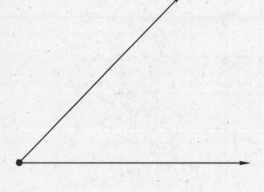

b.

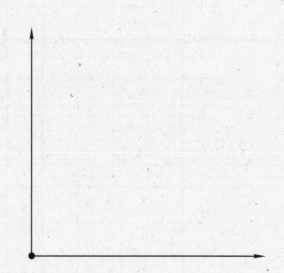

c.

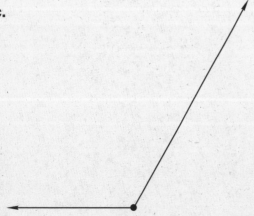

d.

Saxon Math Intermediate 5

Name _____

Cube Pattern

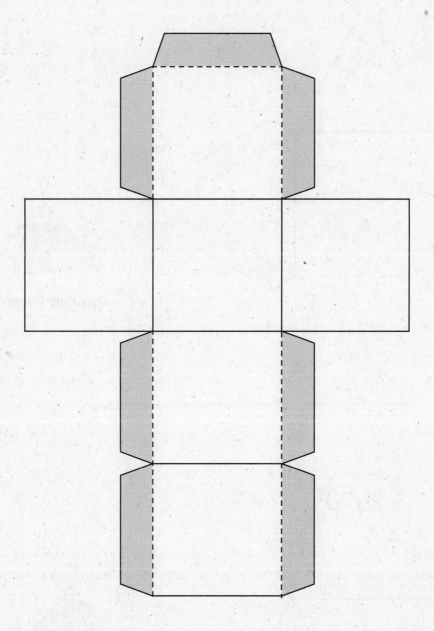

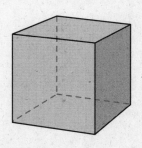

Pyramid Patterns

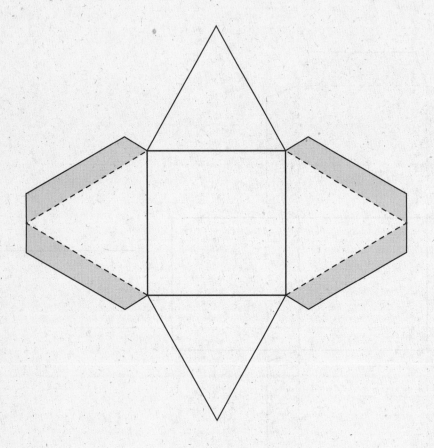

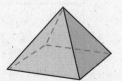

Square Pyramid

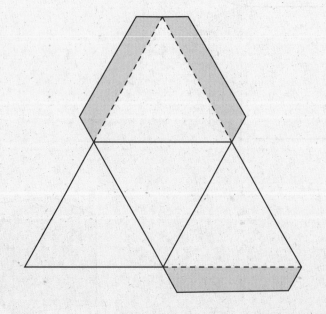

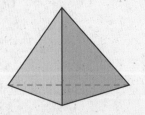

Triangular Pyramid

Saxon Math Intermediate 5

Tessellations, Part 1

Carefully cut out these polygons. Form a tessellation using the triangles.
Then form a tessellation using the quadrilaterals.

Name _____

Tessellations, Part 2

Follow the instructions in Investigation 12 to create a tessellation in the box below.

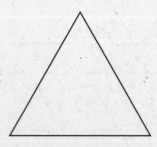

Power Up A

Facts	Add.								
5 +5 10	2 +9 11	4 +5 9	3 +7 10	8 +8 16	2 +6 8	6 +9 15	4 +8 12	2 +4 6	7 +9 16
3 +4 7	7 +8 15	5 +9 14	2 +3 5	4 +9 13	6 +6 12	5 +0 5	3 +8 11	10 +10 20	5 +6 11
0 +0 0	2 +7 9	9 +9 18	5 +7 12	3 +3 6	4 +6 10	2 +2 4	9 +1 10	8 +9 17	3 +6 9
4 +4 8	3 +9 12	2 +5 7	6 +8 14	7 +7 14	3 +5 8	5 +8 13	4 +7 11	2 +8 10	6 +7 13

Power Up B

Facts	Subtract.								
9 −8 1	8 −5 3	16 −9 7	11 −9 2	9 −3 6	12 −4 8	14 −9 5	6 −4 2	16 −8 8	5 −2 3
14 −7 7	20 −10 10	10 −7 3	15 −6 9	13 −7 6	18 −9 9	10 −8 2	7 −3 4	11 −5 6	9 −4 5
12 −6 6	10 −5 5	17 −9 8	13 −8 5	12 −3 9	7 −2 5	14 −8 6	8 −6 2	15 −7 8	13 −9 4
8 −4 4	12 −5 7	9 −2 7	16 −7 9	11 −8 3	6 −3 3	10 −6 4	17 −8 9	10 −10 0	11 −4 7

Power Up C

| Facts | Multiply. |

9 ×6 54	7 ×1 7	9 ×2 18	10 ×10 100	7 ×4 28	6 ×5 30	3 ×2 6	4 ×4 16	8 ×6 48	6 ×3 18
7 ×7 49	4 ×3 12	8 ×5 40	2 ×2 4	9 ×9 81	8 ×3 24	3 ×0 0	9 ×7 63	7 ×2 14	8 ×8 64
5 ×4 20	6 ×2 12	6 ×6 36	7 ×3 21	5 ×5 25	8 ×7 56	3 ×3 9	9 ×8 72	4 ×2 8	0 ×7 0
9 ×4 36	9 ×5 45	8 ×2 16	6 ×4 24	9 ×3 27	5 ×2 10	8 ×4 32	7 ×5 35	5 ×3 15	7 ×6 42

Power Up D

| Facts | Divide. |

7 $7\overline{)49}$	5 $5\overline{)25}$	9 $3\overline{)27}$	8 $3\overline{)24}$	1 $9\overline{)9}$	4 $3\overline{)12}$	4 $4\overline{)16}$	5 $2\overline{)10}$	7 $6\overline{)42}$	7 $4\overline{)28}$
0 $6\overline{)0}$	2 $2\overline{)4}$	7 $5\overline{)35}$	3 $2\overline{)6}$	5 $3\overline{)15}$	9 $6\overline{)54}$	8 $2\overline{)16}$	9 $8\overline{)72}$	6 $5\overline{)30}$	7 $3\overline{)21}$
3 $3\overline{)9}$	9 $9\overline{)81}$	8 $5\overline{)40}$	5 $4\overline{)20}$	8 $7\overline{)56}$	9 $2\overline{)18}$	6 $6\overline{)36}$	7 $8\overline{)56}$	6 $2\overline{)12}$	6 $7\overline{)42}$
8 $6\overline{)48}$	7 $2\overline{)14}$	9 $4\overline{)36}$	6 $4\overline{)24}$	9 $5\overline{)45}$	4 $2\overline{)8}$	6 $3\overline{)18}$	9 $7\overline{)63}$	8 $4\overline{)32}$	8 $8\overline{)64}$

Saxon Math Intermediate 5

Power Up Answers

Power Up E

Facts Divide.

9 9)81	6 8)48	3 6)18	5 8)40	2 3)6	4 7)28	3 5)15	8 9)72	2 7)14	5 5)25
6 9)54	4 8)32	3 4)12	0 4)0	2 6)12	4 4)16	6 7)42	2 2)4	5 9)45	7 8)56
3 8)24	7 9)63	2 4)8	4 5)20	3 3)9	5 7)35	4 9)36	2 8)16	7 7)49	1 8)8
7 6)42	2 9)18	5 6)30	3 7)21	4 6)24	2 5)10	6 6)36	8 8)64	3 9)27	8 7)56

Power Up F

Facts Multiply.

7 ×9 63	4 ×4 16	2 ×5 10	6 ×9 54	5 ×6 30	3 ×8 24	4 ×9 36	2 ×3 6	7 ×8 56	3 ×5 15
5 ×9 45	3 ×4 12	8 ×9 72	2 ×2 4	10 ×10 100	4 ×6 24	6 ×7 42	2 ×8 16	7 ×7 49	8 ×0 0
8 ×8 64	2 ×7 14	3 ×6 18	5 ×8 40	4 ×7 28	3 ×3 9	9 ×9 81	5 ×7 35	2 ×9 18	7 ×1 7
4 ×5 20	6 ×8 48	2 ×4 8	0 ×0 0	3 ×7 21	4 ×8 32	2 ×6 12	5 ×5 25	3 ×9 27	6 ×6 36

Power Up G

Facts Divide. Write each answer with a remainder.

$2\overline{)7}$ 3 R 1	$3\overline{)16}$ 5 R 1	$4\overline{)15}$ 3 R 3	$5\overline{)28}$ 5 R 3	$4\overline{)21}$ 5 R 1
$6\overline{)15}$ 2 R 3	$8\overline{)20}$ 2 R 4	$2\overline{)15}$ 7 R 1	$5\overline{)43}$ 8 R 3	$3\overline{)20}$ 6 R 2
$6\overline{)27}$ 4 R 3	$3\overline{)25}$ 8 R 1	$2\overline{)17}$ 8 R 1	$3\overline{)10}$ 3 R 1	$7\overline{)30}$ 4 R 2
$8\overline{)25}$ 3 R 1	$3\overline{)8}$ 2 R 2	$4\overline{)30}$ 7 R 2	$5\overline{)32}$ 6 R 2	$7\overline{)50}$ 7 R 1

Power Up H

Facts Write these improper fractions as whole or mixed numbers.

$\frac{8}{3} = 2\frac{2}{3}$	$\frac{9}{3} = 3$	$\frac{3}{2} = 1\frac{1}{2}$	$\frac{4}{3} = 1\frac{1}{3}$	$\frac{7}{4} = 1\frac{3}{4}$
$\frac{10}{5} = 2$	$\frac{10}{9} = 1\frac{1}{9}$	$\frac{7}{3} = 2\frac{1}{3}$	$\frac{5}{2} = 2\frac{1}{2}$	$\frac{11}{8} = 1\frac{3}{8}$
$\frac{12}{12} = 1$	$\frac{9}{4} = 2\frac{1}{4}$	$\frac{12}{5} = 2\frac{2}{5}$	$\frac{10}{3} = 3\frac{1}{3}$	$\frac{16}{4} = 4$
$\frac{13}{5} = 2\frac{3}{5}$	$\frac{15}{8} = 1\frac{7}{8}$	$\frac{21}{10} = 2\frac{1}{10}$	$\frac{9}{2} = 4\frac{1}{2}$	$\frac{25}{6} = 4\frac{1}{6}$

Saxon Math Intermediate 5

Power Up I

| **Facts** | Reduce each fraction to lowest terms. | | | |

$\frac{2}{10} = \frac{1}{5}$	$\frac{3}{9} = \frac{1}{3}$	$\frac{2}{4} = \frac{1}{2}$	$\frac{6}{8} = \frac{3}{4}$	$\frac{4}{12} = \frac{1}{3}$
$\frac{6}{9} = \frac{2}{3}$	$\frac{4}{8} = \frac{1}{2}$	$\frac{2}{6} = \frac{1}{3}$	$\frac{3}{6} = \frac{1}{2}$	$\frac{6}{10} = \frac{3}{5}$
$\frac{5}{10} = \frac{1}{2}$	$\frac{3}{12} = \frac{1}{4}$	$\frac{2}{8} = \frac{1}{4}$	$\frac{4}{6} = \frac{2}{3}$	$\frac{50}{100} = \frac{1}{2}$
$\frac{2}{12} = \frac{1}{6}$	$\frac{8}{16} = \frac{1}{2}$	$\frac{9}{12} = \frac{3}{4}$	$\frac{25}{100} = \frac{1}{4}$	$\frac{6}{12} = \frac{1}{2}$

Power Up J

| **Facts** | Simplify. | | | |

$\frac{6}{4} = 1\frac{1}{2}$	$\frac{10}{8} = 1\frac{1}{4}$	$\frac{9}{12} = \frac{3}{4}$	$\frac{12}{9} = 1\frac{1}{3}$	$\frac{12}{10} = 1\frac{1}{5}$
$\frac{12}{8} = 1\frac{1}{2}$	$\frac{8}{6} = 1\frac{1}{3}$	$\frac{10}{4} = 2\frac{1}{2}$	$\frac{8}{20} = \frac{2}{5}$	$\frac{20}{8} = 2\frac{1}{2}$
$\frac{24}{6} = 4$	$\frac{9}{6} = 1\frac{1}{2}$	$\frac{15}{10} = 1\frac{1}{2}$	$\frac{8}{12} = \frac{2}{3}$	$\frac{10}{6} = 1\frac{2}{3}$
$\frac{16}{10} = 1\frac{3}{5}$	$\frac{9}{12} = \frac{3}{4}$	$\frac{15}{6} = 2\frac{1}{2}$	$\frac{10}{20} = \frac{1}{2}$	$\frac{18}{12} = 1\frac{1}{2}$

Power Up K

Facts   Write each percent as a reduced fraction.

$50\% = \frac{1}{2}$	$75\% = \frac{3}{4}$	$20\% = \frac{1}{5}$	$5\% = \frac{1}{20}$
$25\% = \frac{1}{4}$	$1\% = \frac{1}{100}$	$60\% = \frac{3}{5}$	$99\% = \frac{99}{100}$
$10\% = \frac{1}{10}$	$90\% = \frac{9}{10}$	$2\% = \frac{1}{50}$	$30\% = \frac{3}{10}$
$40\% = \frac{2}{5}$	$35\% = \frac{7}{20}$	$80\% = \frac{4}{5}$	$4\% = \frac{1}{25}$

Facts A

Add.

8 $+3$ 11	5 $+6$ 11	2 $+9$ 11	4 $+8$ 12	3 $+9$ 12	6 $+3$ 9	7 $+3$ 10	6 $+4$ 10
5 $+5$ 10	7 $+2$ 9	8 $+5$ 13	2 $+5$ 7	5 $+7$ 12	5 $+4$ 9	2 $+8$ 10	4 $+6$ 10
6 $+5$ 11	4 $+9$ 13	8 $+6$ 14	5 $+8$ 13	7 $+4$ 11	6 $+6$ 12	8 $+2$ 10	2 $+4$ 6
9 $+1$ 10	8 $+8$ 16	2 $+2$ 4	4 $+5$ 9	6 $+2$ 8	5 $+9$ 14	3 $+3$ 6	2 $+7$ 9
4 $+4$ 8	7 $+5$ 12	8 $+7$ 15	3 $+4$ 7	7 $+9$ 16	6 $+7$ 13	9 $+2$ 11	8 $+9$ 17
7 $+6$ 13	6 $+8$ 14	8 $+4$ 12	3 $+5$ 8	9 $+8$ 17	5 $+0$ 5	9 $+3$ 12	2 $+6$ 8
3 $+6$ 9	5 $+2$ 7	6 $+9$ 15	9 $+6$ 15	4 $+3$ 7	9 $+9$ 18	9 $+4$ 13	7 $+7$ 14
3 $+7$ 10	2 $+3$ 5	9 $+5$ 14	3 $+8$ 11	5 $+3$ 8	9 $+7$ 16	7 $+8$ 15	4 $+2$ 6

Facts B

Subtract.

16 -9 $\overline{7}$	11 -3 $\overline{8}$	18 -9 $\overline{9}$	13 -7 $\overline{6}$	11 -5 $\overline{6}$	17 -9 $\overline{8}$	10 -9 $\overline{1}$	13 -4 $\overline{9}$
10 -5 $\overline{5}$	5 -2 $\overline{3}$	12 -6 $\overline{6}$	10 -1 $\overline{9}$	6 -4 $\overline{2}$	7 -2 $\overline{5}$	14 -7 $\overline{7}$	11 -6 $\overline{5}$
16 -7 $\overline{9}$	10 -3 $\overline{7}$	12 -4 $\overline{8}$	11 -7 $\overline{4}$	17 -8 $\overline{9}$	10 -6 $\overline{4}$	9 -5 $\overline{4}$	12 -5 $\overline{7}$
9 -3 $\overline{6}$	12 -3 $\overline{9}$	16 -8 $\overline{8}$	9 -1 $\overline{8}$	15 -6 $\overline{9}$	11 -4 $\overline{7}$	13 -5 $\overline{8}$	8 -5 $\overline{3}$
9 -6 $\overline{3}$	11 -2 $\overline{9}$	10 -8 $\overline{2}$	6 -3 $\overline{3}$	14 -5 $\overline{9}$	8 -6 $\overline{2}$	11 -8 $\overline{3}$	13 -8 $\overline{5}$
7 -4 $\overline{3}$	10 -7 $\overline{3}$	12 -8 $\overline{4}$	8 -7 $\overline{1}$	7 -3 $\overline{4}$	7 -6 $\overline{1}$	7 -5 $\overline{2}$	8 -4 $\overline{4}$
13 -6 $\overline{7}$	15 -8 $\overline{7}$	13 -9 $\overline{4}$	11 -9 $\overline{2}$	10 -4 $\overline{6}$	9 -4 $\overline{5}$	10 -2 $\overline{8}$	8 -3 $\overline{5}$
14 -8 $\overline{6}$	12 -9 $\overline{3}$	9 -8 $\overline{1}$	12 -7 $\overline{5}$	15 -7 $\overline{8}$	14 -9 $\overline{5}$	9 -7 $\overline{2}$	8 -2 $\overline{6}$

Saxon Math Intermediate 5

Facts C

Multiply.

9 × 9 = 81	3 × 5 = 15	8 × 5 = 40	2 × 6 = 12	4 × 7 = 28	7 × 2 = 14	7 × 8 = 56	3 × 4 = 12
5 × 9 = 45	7 × 3 = 21	2 × 7 = 14	6 × 3 = 18	5 × 4 = 20	1 × 0 = 0	9 × 2 = 18	9 × 0 = 0
2 × 8 = 16	6 × 4 = 24	8 × 1 = 8	3 × 3 = 9	4 × 8 = 32	9 × 3 = 27	4 × 9 = 36	8 × 4 = 32
6 × 5 = 30	2 × 9 = 18	9 × 4 = 36	7 × 4 = 28	5 × 8 = 40	4 × 2 = 8	9 × 8 = 72	3 × 6 = 18
5 × 5 = 25	5 × 0 = 0	6 × 6 = 36	7 × 9 = 63	9 × 1 = 9	5 × 1 = 5	4 × 3 = 12	8 × 9 = 72
3 × 7 = 21	9 × 7 = 63	7 × 5 = 35	8 × 8 = 64	8 × 3 = 24	5 × 2 = 10	9 × 5 = 45	6 × 7 = 42
8 × 6 = 48	3 × 8 = 24	7 × 6 = 42	9 × 6 = 54	4 × 4 = 16	5 × 3 = 15	7 × 7 = 49	6 × 2 = 12
4 × 5 = 20	6 × 8 = 48	8 × 7 = 56	4 × 6 = 24	5 × 7 = 35	8 × 2 = 16	6 × 9 = 54	3 × 9 = 27

Facts D
Divide.

3 7)21	5 2)10	7 6)42	3 1)3	6 4)24	6 9)54	3 6)18	6 5)30
8 4)32	7 8)56	2 6)12	6 3)18	8 9)72	3 5)15	6 7)42	6 6)36
2 5)10	3 2)6	9 7)63	4 4)16	6 8)48	2 1)2	7 5)35	7 3)21
9 2)18	5 3)15	5 8)40	4 5)20	3 9)27	5 7)35	5 4)20	7 9)63
4 1)4	2 7)14	3 8)24	4 6)24	8 2)16	8 8)64	7 4)28	7 7)49
2 2)4	9 9)81	4 3)12	5 6)30	4 8)32	4 9)36	9 3)27	7 2)14
5 5)25	8 6)48	4 7)28	9 4)36	9 5)45	2 4)8	2 8)16	8 3)24
5 9)45	9 6)54	8 7)56	9 8)72	8 5)40	3 3)9	2 9)18	3 4)12

Saxon Math Intermediate 5

Test-Day Activity 1

1. Sample: I would use mental math because I can easily add 6 and 2 in my head.

2. Sample: I would use the calculator because I want the exact sum of numbers that aren't rounded.

3. Sample: I would use paper and pencil since I want an exact sum of numbers that aren't rounded.

Test-Day Activity 2

1. See student work

2. Sample: I divided two circles into halves. I counted the halves. There were 4 of them.

3. 6; See student work; Sample: I divided the drawing of $\frac{3}{4}$ of a circle into eighths of a circle. I counted the eights. There were 6 of them.

Test-Day Activity 3

1. Sample: Triangle *GHI* is congruent to triangle *MNO* because they are the same size and shape.

2. **a.** vertex *H*; **b.** side *NO*; **c.** angle *G*

3. Sample: Angle *A* corresponds to Angle *F*. Side *BC* corresponds to side *GH*.

Test-Day Activity 4

1. See student work; $\frac{1}{2} \times \frac{1}{4} = \frac{1}{8}$

2. See student work; $\frac{1}{2} \times \frac{3}{4} = \frac{3}{8}$

3. Sample: I could look for a fraction piece that is half the size of $\frac{1}{4}$. I would find $\frac{1}{8}$.

Test-Day Activity 5

1. $\frac{2}{4}$

2. $\frac{4}{12}$

3. $\frac{6}{8}$

4. $\frac{3}{6}$ and $\frac{4}{6}$

5. $\frac{4}{6}$ and $\frac{5}{6}$

6. $\frac{3}{6}$ and $\frac{1}{6}$

7. $\frac{6}{12}, \frac{4}{12}, \frac{3}{12}$

Test-Day Activity 6

1. See student work

2. See student work

3. Sample: Both graphs are good ways to show the information. An advantage to the bar graph is that it shows individual numbers that can be compared. An advantage to the circle graph is that it makes it easy to see the whole and what part of the whole each toy robot section takes up.

Test-Day Activity 7

1. See student work

2. See student work

Test-Day Activity 8

1. Sample: Area = 2 in. × 6 in. = 12 sq.in.; Area = 3 in. × 3 in. = 9 sq.in.; Total Area = 21 sq.in.

2. Area of rectangle = 3 cm × 4 cm = 12 sq. cm;
   Area of triangle = $\frac{1}{2}$ × 2 cm × 3 cm = 3 sq. cm; Total Area = 15 sq.cm

Test-Day Activity 9

1. Sample: There are about 21 whole square centimeters in the circle. If you put all the small parts of squares that are in the circle together, there is about 1 more square centimeter. So the area of this circle is about 22 square centimeters.

2. 14 squares, which is 70 square feet; Sample: when you look at the figure on a square grid, you can count the number of whole squares that are in the circle. Then you can estimate how many more whole squares there would be if you put all the leftover parts of squares together.

Test-Day Activity 10

1. Cubic yard

2. Cubic meter

3. **A.** Sample: Centimeters are lengths; I put a pencil beside a centimeter ruler to measure it.

4. **B.** Sample: Square centimeters cover area.

Test-Day Activity 11

1. 15 square feet

2. 24 square inches

3. 12 sq. mm; 0.12 sq. cm

4. See student work; Sample: Twelve squares are inside the rectangle, which is 12 square millimeters. Twelve squares is 12 of the hundred parts of a square centimeter, so the area is $\frac{12}{100}$ or 0.12 square centimeters